I Live, I Fight, I Win

Rules of Life In War

ALL-RUSSIAN NON-GOVERNMENTAL ORGANIZATION UNION OF VETERANS OF AFGHANISTAN AND SPECIAL MILITARY OPERATIONS

Adapted North American Edition
Carentan Media Group
Move * Shoot * Communicate

Printed in the United States of America

Adapted Edition 2024

ISBN xxxxxxxx

Carentan Media Group

Carentanmedia@proton.me

Introduction

This work is an open-source document procured from Russian sources. It has been translated openly by various organizations using Google Translate. We have edited for grammar and added a photo section.

We present this in printed for informational and historical purposes.

The war in Ukraine started in 2014 with the violent annexation of the Eastern Ukrainian territories of Donbass and Luhansk, as well as Crimea. The Ukrainian army at the time was poorly equipped, poorly trained, and badly led, the byproduct of 30 years of corruption and post-Soviet incompetence. What followed was a fixed line stalemate that lasted until Russia attempted to fully invade the country on Feb. 24th, 2022.

By all accounts it was expected to be a swift Russian victory. Over the years of the Trump administration, however, lethal weapons aid, notably in the anti-tank arena, and Russian incompetence meant that key cities of Kiev, Kharkiv were held by the UA. Along the coast the Russians were able to secure the port city of Mariupol. After a Stalingrad-like stand by UA and Ukrainian marines.

The conflict has since stabilized into a WW1-esque trench warfare stalemate with gains measured in yards and casualties in the hundreds of thousands. Unlike WW1, however, the battlespace is infested with modern technology that has made survival on the battlefield much more difficult.

It continues as of this writing in March of 2024.

** Carentan Media Group in NO WAY endorses the rationale for the war outlined herein by the Russians who wrote it.*

I Live, I Fight, I Win: Rules of Life In War

ALL-RUSSIAN NON-GOVERNMENTAL ORGANIZATION UNION OF VETERANS OF AFGHANISTAN AND SPECIAL MILITARY OPERATIONS

(Original appendices were absent at time of translation)

I Live, I Fight, I Win : Rules of Life In War
Razumov A.N., Kryukov G.A., Kuznetsov A.N.

I live, I fight, I win! Rules of Life in War is a collection of recommendations based on the generalized experience of combatants in Afghanistan, the North Caucasus and Ukraine. The presentation is distinguished by a deep knowledge of the problem, brevity, accessibility for understanding, clarity of presentation.
The collection is intended for servicemen of the RF Armed Forces participating in a special military operation in Ukraine, conscripts, cadets of military educational institutions, employees of various law enforcement agencies.

Because each war is unique in terms of its methods of waging, the nature of the theater of operations and the bitterness of the belligerents, there is not and cannot be a single reference book for all cases of military activity. But military life, of which 80% of any war consists, is the same everywhere. Ignorance of its basics undermines the health of a fighter and leads to losses. Therefore, veterans who fought in Afghanistan, the Caucasus and Ukraine did not describe in this brochure the features of fighting in the city, at sea or in space. This manual will be taught by commanders and senior comrades. We simply summarized our experience of life in the war, collected the main thing that helps to survive and win.

Franz KLINTSEVICH
Leader of the Russian Union of Afghan Veterans and Special Military Operations

Important! Got the agenda. What to bring to the recruiting station
Do not panic. Do not believe the rumors and fools who claim that everything from a helmet to body armor must be purchased independently. Do not burden your family with expenses and unnecessary expenses.
You will receive everything you need at the place of stay. It is better to come to the collection point in a tracksuit and sneakers (sports shoes).

With me I have:
- passport, military ID, driver's license;
- toiletries and shaving accessories for 1 month;
- adhesive plaster 2–3 packages;
- loperamide (anti-diarrheal);
- folding knife with an awl (multi-tool), spoon;
- nippers for nails (knipser);
-3-4 pairs of socks;
- 2-3 sets of underwear (loose according to cut);
- 3–4 pairs of work gloves;
- a set of needles with threads (green, black, white; a lighter, matches (preferably all-weather)
- a block of cigarettes (for smokers);
- laces for berets (110–120 cm);

1. What is a special military operation

The politicians called the actions of our army in Ukraine a special military operation. From the point of view of international law, this is true. And for its participants, this is a real war with blood, pain, bitterness of loss and joy of victory.

2. Great Patriotic War 2.0.

Just look at the list of countries that have declared sanctions and are helping the Ukrainian regime—Germany, Poland, Czech Republic, Croatia, Norway, Denmark, Japan, Italy... They all fought against us. Today in Ukraine they are taking revenge on Russia for our Great Victory. So, for us, this is a continuation of the Great Patriotic War. And we, like our grandfathers in forty-five, are obliged to win.

3. In Ukraine, we protect Russia

There is no Ukraine as a state, there is the territory of the former Ukrainian SSR, temporarily occupied by a terrorist gang. All power there is concentrated in the hands of the citizens of Israel, the USA, and Great Britain, who organized the genocide of the natives, "reducing" 20 million people during the years of "independence". To survive, people, like the Negroes in the United States, once worked for pennies on European plantations. The brothels of Europe are filled with young Ukrainian women. Men are forced to fight against Russia. The United States and Europe want to wage this war to the last Ukrainian. They don't need people. They need territory and resources. They prepared the same future for us. Therefore, fighting in Ukraine, we are defending Russia and saving residents of Ukraine from the genocide unleashed by Ukrainian and Western politicians.

4. Who are Ukrainians

More recently, 96.7% of Ukrainians were Russians. But for 30 years of independence, they were deprived of a normal education, culture, native language and turned into "wild" Russophobes. They still have something left of the Russians. They, like us, were brought up on the exploits of their grandfathers who defeated fascism. They are the same brave fighters - staunch on the defensive, daring on the offensive. Someday, after denazification, they will become Russian again, but for now they are enemies. Cruel and insidious. This means that they must be beaten until they lift their hands up, without relaxing, until our very victory. (Ukrainians apparently disagree with this)

5. Features of the strategy of the Ukrainian army

The West has been preparing Ukraine for war with Russia for 30 years. Realizing that we cannot be defeated in a fair fight, they chose the meanest strategy. Following the example of the Nazis, they organized a focal defense system in settlements. There they deploy battalion or company tactical groups reinforced with artillery, sabotage units and modern reconnaissance equipment. The strongholds are equipped with cable internet, which allows online surveillance, transmission of commands and inflicting fire damage. In addition to Soviet weapons,

Ukrainian units are equipped with NATO ATGMs, rocket and artillery systems. Each platoon has quadcopters, thermal imagers and protected means.

6. Ukrainian tactics - "Hit and run."

When Russian troops are detected, it is enough for the Ukrainian military to poke a finger at the screen of the tablet, so that the coordinates of the target enter the unified battle management network, and the free artillery unit receives a command to open fire. Artillery, located in a shelter (workshops of factories, basements of shopping centers, etc.), advances to previously prepared positions, strikes, and returns to their shelters. Mobile sabotage groups complete the defeat of the enemy, which, after a short fire contact, return to their strongholds, hiding behind civilians.

7. Combat losses and not so

The nature of the fighting in Ukraine is different from Afghanistan and Chechnya. In terms of the saturation of the parties with heavy weapons, they are closer to the Second World War. This is evidenced by the loss statistics. Among the dead, up to 78% have shrapnel injuries, 12% - bullet wounds, 10% - mine-explosive wounds. And now the important thing! In war, up to 40% are non-combat losses.

Causes; violation of safety measures when dealing with weapons, defeat from "friendly fire" due to lack of interaction with neighbors, transport accidents, hazing, illness, poisoning with alcohol substitutes. The Americans were unable to change the statistics of the causes of non-combat losses. But they almost doubled their number to 23%, obliging military personnel in the combat zone to constantly wear a helmet and body armor.

Conclusion 1: Helmet and body armor doubles the chance to survive.
Conclusion 2: Discipline and safety measures help prevent getting into the statistics of non-combat losses.

8. Where we lose and how not to become a loss

The success of the enemy is, most often, the result of our mistakes. We suffer the main damage from artillery fire. On the march, in places of concentration when setting tasks, when organizing a field camp or stronghold, as well as the actions of sabotage groups. In order not to replenish the statistics of losses, you should follow simple rules.

1) On the march or at a halt, keep a distance between vehicles of at least 20-30 meters.

This reduces mass casualties from precision weapons by 90%. Any boss who places equipment closer than 20 meters from another is a fool and a pest.

2) Dig in even in places where units are temporarily located.

Stopped - dig in. A small trench is a good cover from shrapnel and shock waves. (This is old Soviet doctrine from WW2 that served them well)

3) Be constantly on the lookout.

Wherever you are - on the front line or in the deep rear - monitor the sky and the surrounding area.

4) Organize yourself.

The crew of each vehicle (squad) must become an independent tactical unit, providing protection, camouflage, maintenance of military equipment, engineering equipment of positions, meals and rest for soldiers.

9. We think for three

Minimizing losses allows the creation of autonomous combat groups within the unit. The combat trio, formed on the basis of friendly relations, is the best order of self- organization of servicemen. It is formed in the squad around the collective weapons of the unit - a machine gun, a grenade launcher, etc., making up their calculation. Such a distribution increases the effectiveness of heavy weapons and leads to a reduction in losses. If one of the fighters is wounded, the other two are able to quickly evacuate him and provide first aid. The engineering equipment of a firing position for three takes half as much time as digging an individual trench. At the same time, while two people work with shovels, the third one can observe, provide camouflage of the object, cook food, clean weapons, etc. At night, two are alternately awake and watching, one is resting.

10. Memo to the commander

Commander, remember: your task is to manage the battle and, by the way, the relations within the unit. And to become a good commander, you need to kill the militant in yourself and restore the decider. To begin with, throw out the ammunition from the satchel. Leave two magazines for AKs and a couple of efoks(?). This is enough to stand up for yourself and deprive the temptation to shoot enemies "personally".

What a commander really needs to have is a set of spare batteries for the radio station, good binoculars, a GPS navigator, a laser range finder, a thermal imaging monocular and chargers for them.

The Ministry of Defense does not provide these gadgets, so the commander purchases them on his own. It's time to get used to the fact that the commander, like any professional, must have personal working tools.

Provide your fighters with everything necessary, organize guards, equip positions, train subordinates in the use of all types of weapons. Then the commander in battle will not have to personally sit in the gunner's seat or mechanics.

Don't let anyone ignore you. Take away the machine gun from the soldier who defies you in response to a fair demand for service. Let him fight with a shovel until he grows wiser. The chief, who thwarted the delivery of food, ammunition, fuel to the unit, was using all available means: from complaints to the higher command and statements to the military prosecutor's office, to hazing. He is a traitor and an enemy. With its inaction, it destroys you and the soldiers entrusted to you.

Firmly demand clarification of the mission, providing communications and artillery support. Any refusals are ignoring you as a person and as a commander. And if you are dynamized by the bosses, then the fighters will also ignore you. Don't stoop to the level of "Terpila" (A slang term for a weakling or pussy. Someone who takes any insult or beating without retaliation) Be the leader! Only in this way will you bring order to the unit and save your fighters.

11. Soldier's truth

In Afghanistan, the fighters of the Kunduz reconnaissance battalion had a tradition. On operations, they did not start eating until the commander took the spoon. True, this did not apply to everyone, but only to those commanders who, with their professionalism, ensured the fulfillment of combat missions and the preservation of subordinates. A good commander is a master of his craft. He constantly teaches fighters how to use weapons. In its positions, trenches of a full profile, observers are always at their posts, rest and food for personnel are organized according to allowances. He does not drink alcohol himself and does not allow relaxation among his subordinates. He is calm in battle. Controls the fire of the unit, but, if necessary, skillfully fires from any type of weapon. He keeps steady with superiors, is not afraid to stand up for subordinates, and is not lazy to fill out award lists for those who stand out.

Such a commander must be protected in battle and assisted in his service. To speak honestly with him about problems, to put buzzers and loafers in their place, to drive out cowards. If the commander "relaxes" bitterly, does not delve into the problems of providing and living for the personnel, does not engage in combat training, cannot organize service, and the personnel is left to itself and wanders around the district in search of adventure, then this is a bad commander. It is necessary to leave him as soon as possible, because he and his unit are doomed to unjustified losses.

12. Rule of survival number 1. Don't stand out.

The fighting in Ukraine has become a benefit of drones and all kinds of reconnaissance devices. The location of the unit is calculated from the movement of military personnel, smoke and reflections of lines, heaps of debris in positions - everything that looks unnatural and stands out on the ground. According to the same principle, snipers and gunners of heavy weapons choose the target.

To avoid becoming a target the main principle of disguise must be observed – do not stand out. You should not show off and stand out in uniform among colleagues. New or sophisticated equipment will definitely attract the attention of the enemy, who will take you for a commander or a tough commando. Therefore, before putting on a new uniform, it must be thoroughly washed several times with a brush. After that, it will acquire a softness that is pleasant to the body and a color that blurs the contours of the figure in any terrain.

Another unmasking sign is your shadow. It is better to walk and move along the shady side of the street so that your silhouette and the cast shadow do not contrast with the general background. In the forest or plantings, you need to move at a distance from the edge, behind the second or third row of trees.

13. Don't hang around

The main unmasking sign on the front line is movement. A soldier sitting in a covered trench or trench like a fox hole and even lying motionless in a firing position is not visible to the enemy. He becomes a target as soon as he begins to rummage around the position. If you have to move to the front, then do not wave your arms, keep your elbows, hands and weapons close to the body. This helps to hide the contours of the figure. And further, the active movement of cars and people in uniform betrays the location of the headquarters, warehouse, canteens, etc - places of accumulation of personnel. And this is a priority target for enemy artillery. Try to avoid such places near the front and unnecessarily there don't fidget.

14. Bored? Wait, it's coming.

Don't bunch up. The desire of people in moments of danger to stay in a group makes you a group target. The movement of even a small group in a column one at a time with an interval of 4–6 steps arouses the interest of the reconnaissance drone operator and is guaranteed to end with an artillery arrival. Afghan Mujahideen, in order to avoid such a defeat, moved in pairs with an interval of up to 100-150 meters. Groups of 2 people are not as conspicuous as a squad or platoon and does not represent great interest as a target for artillery.

15. We do not need such a movie

In addition to unmanned and radar reconnaissance equipment, our enemy is actively using video surveillance equipment - from cameras on the roads to smartphones with schoolchildren. This "movie" in real time delivers intelligence information to the firing positions of the enemy. In order not to become a "movie hero" in someone else's sight, it is necessary to constantly monitor the area around your location.

First, check the places where there may be observers or security cameras. As a rule, these are hills dominating the area, factory chimneys, buildings that offer a view of your location, poles at intersections or road sections with maximum visibility, and power transmission towers. Detected observers or suspicious sites are immediately destroyed by all available means of destruction. With CCTV Cameras it is enough to de-energize by interrupting the electric supply.

As for “Shkolota” (slang term for a snoop, or nosy person. Literally school child. Here it refers to civilians recording Russian troops) with smartphones, for starters, you can ask for a camera to watch and “accidentally” drop it on the asphalt. After 5–6 such “views” of the shooting stop.

16. If the locals are greyhounds

All protest actions, sabotage, espionage, and sabotage by the local population in the liberated territories are carried out from the command centers of the Armed Forces of Ukraine and the SBU. This means that, in accordance with the First Additional Protocol to the Geneva Conventions of 1978, article 45, protesters can be considered combatants who do not distinguish themselves with uniforms, insignia, etc. (The above article does not say this. Civilians participating in armed conflict are by law to be treated as prisoners of war.)

This allows you to detain local residents, hinder the fulfillment of their official duties by our servicemen, and treat them as prisoners of war, and in the case when the locals pose a threat to the life and health of the fighters, open fire on them. Article 46 of the same First Additional Protocol to the Fourth Geneva Convention allows us to consider all local people who take photos and videos of our positions as potential spies. And this, in the conditions of hostilities, deprives them even of the right to be considered prisoners of war with all the ensuing consequences ...

But! Shooting at the unarmed is not our method. There are dozens of ways to indicate your status quo. Protesters and those who are filming our objects, for example: You can detain them like prisoners of war, carry out investigative actions with them, send them to the disposal of the competent authorities. The main thing is to act quickly, decisively and with the ingenuity characteristic of our people.

17. An idiot with a cell phone is his own enemy.

Any turning on of a mobile phone in a war zone leads to the detection of the subscriber with the establishment of his geolocation and subsequent artillery strike by location. The fact is that when used in a smartphone, a SIM card automatically connects to a nearby tower and the Ukrainian cellular operator, seeing your Russian number, finds it, determining its location with an accuracy of 3 meters.

Moreover. When detecting your phone number, the Ukrainian mobile operator just needs to enter it into the NT-Locator application and start the search, after which not only the current geolocation and movement data become available, but also information about the subscriber himself, including the last name and home address.

Therefore, do not be surprised if, after a phone call from the front line, your position is shelled, and your relatives are called with threats, insults and extortion of money.

18. Photos in social networks are a godsend for a spy

The location of your smartphone can also be determined by geotags - data in social networks for sharing photos and video.

"If a user specified a location when publishing photo and video content, then any other user automatically gets access not only to the image, but also to the place where the picture was taken."

The problem is that in the image information (EXIF data), smartphones with a GPS module indicate the date of shooting, day and exact time, country, city, address, GPS coordinates of latitude and longitude. What if the picture was taken inside the building, then also the exact address with the street name, house number and even the postal code. The Ukrainian operator only has to upload your photos to the EXIF data viewing service and send you a 152 mm package or a HIMARS package.

19. How dangerous is artillery fire

The Armed Forces of Ukraine are armed with Soviet and foreign systems of cannon, rocket artillery and mortars close to them in terms of characteristics. A cannon artillery shell and a mine make a sound in flight, reminiscent of the flapping of the wings of a dove. The sound of a flying rocket projectile is reminiscent of the sound of slate being pulled over sand. The sound of a detonation is sharp, short, with a characteristic crack of tearing metal. A black cloud of a mixture of dust and burnt explosive up to 2–3 meters high forms at the site of the projectile explosion. AT mines - up to a meter. When undermining the most common in the Armed Forces of Ukraine - 122 mm high-explosive fragmentation projectile, 400-500 lethal (weighing at least 5 grams) fragments are formed. Half of them go into the ground and fly out of the affected area, losing their lethal force. The effective radius of destruction on solid ground is 30 meters. On soft soils, when a funnel is formed 1 meter deep and 2 meters wide, the damage radius is reduced to 10 meters. The spread of fragments reaches 400 meters. Affecting factors 152 mm projectile 1.5 times above.

An 82 mm mortar fragmentation shell, when detonated, produces 150–200 lethal fragments with a damage radius of at least 18m. The spread of individual fragments can reach 100– 150 meters. The diameter of the funnel at break is up to 80 cm. depth - about 10 cm. The damaging factors of a 120 mm mortar mine are 1.5 times higher.

20. Artillery is not as terrible as they say about it

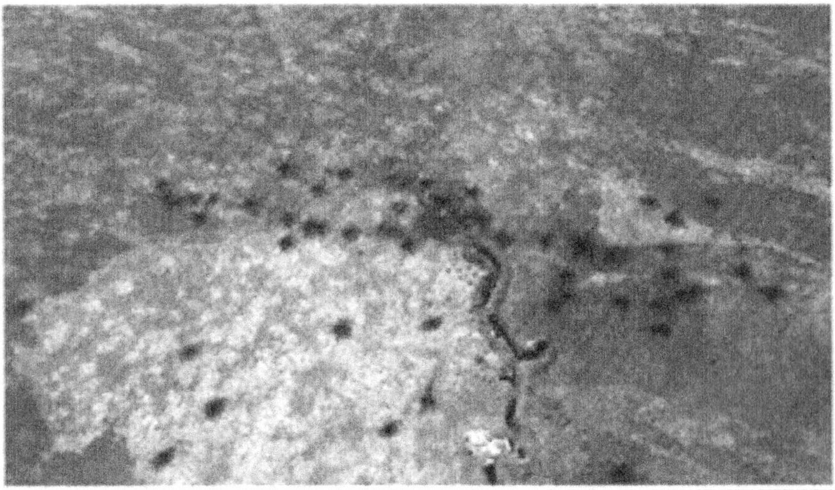

Image 1: The results of artillery shelling of the position of the squad. A big lay down, but all past. Not a single soldier in the trench was hurt.

Calculations show that only 160 shells of 122 mm caliber, or 8-10 per soldier, are required to defeat a platoon in open areas. But to defeat 75% of the targets in an equipped platoon stronghold, 1,250 high-explosive fragmentation shells are already required, which is 60 shells per fighter. Taking into account the shipping container, this is almost 2.5 tons, which still need to be delivered to the guns, unloaded, and then loaded into the barrel.

The conclusion is simple. An artillery strike is not as terrible as the reaction of untrained servicemen to it. Even a primitive prone firing trench or a shell crater reduces the risk of being killed or injured by 10 times. A steel helmet and body armor protect a person's vital organs from most fragments. The armor of armored personnel carriers, infantry fighting vehicles, fragments of shells and mines do not penetrate. And further. It takes a whole 5 minutes for a howitzer battery to fire 160 shells at an identified target, and we have enough time to take cover from artillery fire or leave the affected area.

21. Signs of an impending artillery raid

The appearance of unmanned aerial vehicles in the sky above the unit may be a harbinger of an imminent artillery strike. The explosion of a smoke projectile (white smoke) near our positions means artillery sighting - the determination of parameters for shooting to kill. Paired bursts of shells at our positions, one with overshoot, the other with undershoot. The enemy uses a fork - a technique for capturing a target in range. Urgently take cover or change position.

22. It is difficult to prevent an artillery attack, but it is possible

As a rule, the artillery attack is carried out on a pre- identified target, so do not become a target, disguise yourself. Fire correction indicates the presence of observers or technical means of artillery reconnaissance (drones, quadrocopters, cameras for fixing roads, etc.). Quadcopters can be destroyed centrally by the fire of the unit, even from rifle weapons. With light and medium UAVs, it is more difficult to control the front line of defense. We need funds here for EW and air defense.

Artillery spotters can be within line-of-sight of objects dominating the terrain, from which a good view of the positions is opened. Their presence is established by observers and suppressed by regular fire.

23. During an artillery raid, the main thing is calmness

Ukrainian artillerymen usually fire in "series ": 2-3 shots, 3-4 minutes pause to make adjustments, then again 4-6 shots to finish off. As a rule, there are no more than three such series. If the artillery fire is rapid with a high consumption of ammunition, then our artillery is working and we must urgently establish cooperation to stop this "friendly fire". At the first sign of shelling, immediately fall to the ground or dive into cover (under armor, into a trench, a recess in the ground, a shell crater, a capital structure).

Do not try to hide from the artillery attack under the trees . The projectile fuse (mine) will work when it hits a branch and forms an air blast, increasing the affected area. Crawling movement. No running and panic. If you come under artillery and mortar fire during the march, then leave the affected area at full speed. It is difficult for enemy artillerymen to calculate and conduct mobile fire. Stopping this, you turn into an ideal, motionless target. Help the wounded after the shelling and in shelter, otherwise you will need help already.

24. This cat has 9 lives, and the sapper has one!
The second reason for our losses after the artillery of the Armed Forces of Ukraine was mines. There are several rules in the work of detecting and disarming them:

1. Do not dismantle ammunition, do not experiment with it, and do not try to neutralize explosive devices found.

2. Watch where you're going.
First, study the route at eye level - if there is anywhere a fishing line or wire stretched to an explosive device. Then look down at your feet. Of course, it is difficult to see the PMN hidden in the ground, but the PFM-1 "Petal" mine is quite distinguishable.

3. Detour of the destroyed sections of the road along the sides can be mined.

4. There will most likely be another 2-3 mines.
(Mines are almost always laid in clusters)

5. The corpses of our fighters, on the territory of the enemy, are most likely mined for unloading.
Before evacuation, they are moved from their place with a cable or ropes.

6. The first question that is asked to the prisoner is where the mines are planted in the defense system of his unit.

7. Tripwire grenades will not protect you from enemy DRGs. Specialists will remove them without problems, but their own servicemen are guaranteed to be blown up.

25. Military traffic rules

To avoid losses on the march, you must follow simple rules: the distance between cars is at least 30 meters. This will exclude the possibility of hitting the columns with concentrated artillery fire. To repel attacks on the flank of the column from behind the garden, the barrels of regular weapons of military equipment (tanks, infantry fighting vehicles, armored personnel carriers) are deployed in a "herringbone". Barrel is traversed to the right, at an angle of 45 degrees in the direction of travel, odd numbered vehicles to the right, even - to the left. This provides immediate concentration of fire to destroy the enemy. Personnel are inside the armored personnel carrier (BMP), and not riding on top. The armor serves as a reliable shelter from bullets and fragments, and the absence of crowding on the armor excludes, in case of confusion, losses from one's own fire. Observation over the armor, hiding up to the waist in the hatches, is carried out by the commander of the vehicle (in the direction of movement) and two observers (on both sides of the road). They are also involved in repelling fire with standard weapons. The experience of military operations in Afghanistan and the North Caucasus has shown that a single movement (even of combat vehicles) increases the risk of their attack by enemy DRGs.

26. The night is dark and full of horrors

The advent of night vision sights (NVGs), including thermal imaging ones, gave rise to a lot of rumors about their exceptional effectiveness. In fact, all night devices are more useful as surveillance devices, but for effective fire they are not very good. Too many conditions must be met in order to successfully apply such a night light. Among them are the possibility of its zero on weapons, the state of the batteries, combat factors affecting operation, etc. The constant work on night vision devices tires and blinds the observer and makes him a target himself. The pupil of the NVD eyepiece illuminates the observer with a greenish glow, easily observed through other devices. Therefore, it is necessary to turn on NVD only when necessary. To protect yourself from an enemy with night vision devices, you must, firstly, observe blackout protocols: Do not light a fire or smoke. In the "night light" the light of a cigarette is noticeable at a distance of up to 500 meters, a burning match - up to a kilometer, the light from a fire - up to 8 kilometers.

Second, don't make noise. Human steps on soft ground can be heard up to 30 meters. Coughing and speaking - up to 100 meters. The scream and noise of a running diesel engine - up to 1000 meters.

Thirdly, it is necessary to choose low or closed places for moving at night, to avoid light clearings. Move from cover to cover. And further. Night vision devices are damaged or neutralized by bright light sources. A bright fire in the neutral zone, illuminating reactive cartridges launched along a flat trajectory towards the enemy, force him to stop observing and protect the lens of his device.

27. Bullet is a fool

In combat operations in Ukraine, losses from small arms fire amount to 12%. Less than half of them (or 5% of the total casualty statistics) are dead. Such low effectiveness of small arms fire is characteristic of all recent wars. For one killed in the Great Patriotic War, there were about 25,000 cartridges fired, in Vietnam - already 250,000. And this is 2.5 tons of ammunition by weight. The low efficiency is explained by the psychological unpreparedness of the soldiers for a real battle. Most often they fire in the direction of the enemy without aiming, from the animal the one from the chest above the head, while “burning” an incredible number of rounds.

28. Bac-bac... and over

More about marksmanship. The range of effective shooting from modern types of small arms from an unprepared position does not exceed 150 meters, and the maximum hit is recorded at a distance of up to 75 meters. Let's take into account the features of automatic fire, characterized not only by an increased consumption of ammunition, but most importantly, by a large spread of bullets. If at 100 meters the first bullet of the line from the machine hits the center of the belt target, then the second one flies 60–70 cm higher and to the right. The third and fourth bullets pass already 1.5–2 meters from the target. Shooting combat, of course, is not for the weak, but you should not be afraid of it either. Irretrievable losses in such a battle, during the Afghan war, for example, turned out to be less than 0,05% of the total number of Soviet warriors.

29. Attention! Sniper at work.

So, on average, it takes half a "Ural" (Russian truck) of rounds to bring down a fighter. A good sniper spends no more than two bullets on hitting a target. For this, snipers are not loved either by their own or by strangers. They are a headache for everyone. The enemy is kept in suspense by sniper fire, their own lives in anticipation of a mandatory responsibility for his deeds.

Signs of a sniper's work:

- any bullet hit in the head.
- hit against the background of sound masking - artillery or grenade launcher shot, etc.
- selective bullet damage to an officer, signalman, observer.
- single shot, into the wall near your position on any noticeable spot. This is with a shooter at a specific landmark with a range.
- shooting a fighter walking last in the group so that those in front do not understand what is happening;
- non-fatal wound of a fighter with subsequent defeat of all those who assist him.
- shooting the curious, caught on the bait in the form of specially planted nishtyakov (slang for something awesome, a goodie, a prized item) — canned food, cigarettes, bottles, etc.
- appearance of a quadcopter over your position during shelling. The second number of the sniper pair is working with him. He corrects the fire.

30. Where is the sniper hiding?

In broken armored vehicles. Fire and smoke remain inside, the sound is muted. At the edge of the forest (landing) behind small shrubs or hanging down branches of trees so that the flash and the smoke from the shot did not go beyond these branches. From the depths of the premises located on the shady side of the building. In this case, the arrow is almost invisible, and in order to hide the flash of the shot, the snipers hide behind furniture. In the window opening of the building, the sniper should be looked for in the lower right (away from you) corner behind the natural cover for him (window corner, window sill, wall) that will to his left. In a trench under a fallen or specially placed concrete slab, covering from stray bullets and splinters. Previously, a sniper often used an artificial stump, modern shooters use an old tire. The embrasure in it is disguised as a natural feature.

31. Fight with enemy sniper

It all comes down to finding a sniper position and pouring everything that flies, burns, and explodes there.

1. Calculation of the sector in which the sniper works:

The most prompt is the report of the observer who spotted the shot. If they missed it, then the survey of witnesses of the bullet defeat of our fighter, the study of the trajectory of the arrival of the bullet when hit, are used; A sniper is issued by his support group (from 1–2 to 5–7 people) and quadrocopters of an assistant sniper. The drone flies no more than 20-30 minutes and returns as operator rule.

2. Determination of the distance to the sniper's position:

After determining the sector of the sniper's work, based on data of the effective distance of bullet fire; a hit in the head - 50–100 meters, in the body - up to 300 meters, harassment fire or a "terror" (shooting at everything that protrudes) -up to 500-600 meters.

3. Sniper suppression

The best solution is massive shelling from heavy guns, mortars or MLRS in order to "plow " the detection square. Very effective fire from regular tanks, infantry fighting vehicles, armored personnel carriers.

Volley fire from RPGs and grenade launchers or such exotic things as counter-sniper units will also do. If they are there, of course.

4. Anti-sniper prevention.

Having found any suspicious position close to one's positions, suitable for hiding an enemy sniper (an embrasure in a wall, a gap, etc.), one should bombard it with the fire of tank guns, infantry fighting vehicles, armored personnel carriers, RPGs and regularly fire from small arms, creating a threat to enemy sniper, especially before our offensive.

32. How to behave in places where a sniper works

Don't come out of hiding. Any movement is only running from cover to cover. Identify places where terrain features favor the work of an enemy sniper and put signs there with the inscription: "Attention! Enemy sniper at work!" It saved many lives during the war years.

Those wounded by a sniper are evacuated from an open place in armored vehicles (if there are no risks of its defeat from anti-tank weapons) or by dragging to a shelter with a rope and climbing carabiner. If you need to drag a load across an open area of fire, do it lightly by running or crawling through a dangerous place, and then pull with a rope.

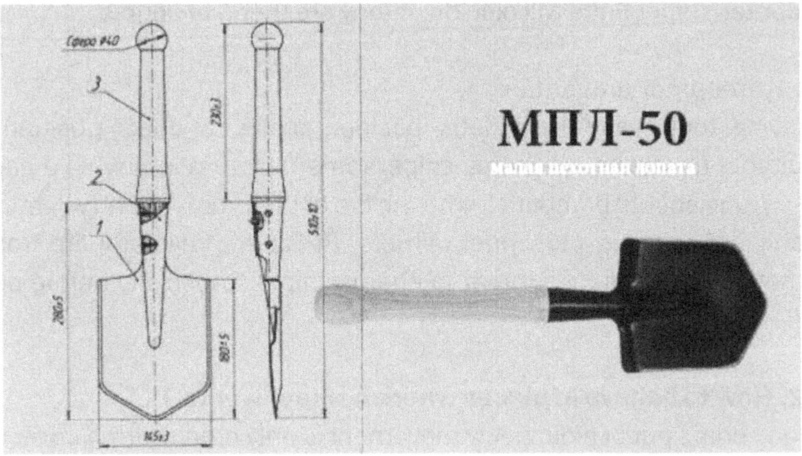

Standard issue Russian military shovel. Changed little for 100yrs or more.

For scale. Individual soldier digs with the implement. Impressive how many hundreds of miles of trenches can be dug with these things.

Miserable conditions in a front-line outpost trench. Note grenades and AKM readily available.

View from inside dugout or covered way. Note camouflage netting and chain link “door” to thwart drone incursions.

Despite flash hiders, night firing is easily detectable.

Small commercial drone with improvised explosive payload. "Kamikaze" drones have added a sinister new dimension to an already miserable situation.

33. Don't snipe

Even if you are an excellent shooter and operate as part of a unit, do not conduct the so-called "hunt" from your location. The enemy will soon get tired of your shooting and an artillery or mortar strike of retaliation will surely fall on the heads of your colleagues. Operate either as part of your unit, or from standalone positions.

34. Personal weapon

The choice of personal weapons for a participant in the special military operation in Ukraine is small. Most often these are AKM, AK-74 and their modifications. 7.62 mm AKM is heavier. Thanks to a powerful cartridge, it is not sensitive to dirt and powder deposits. The 5.45mm AK-74 is lighter. It's more comfortable to shoot. If you still have a choice, then it is optimal to take weapons for the ammunition used in your unit.

35. Problems of accuracy of shooting from AK

Properly brought to normal combat, a Kalashnikov assault rifle allows you to confidently hit the handle of an aluminum spoon at 100 m. As for the complaints about the low efficiency of automatic fire, the main problem is the psychological unpreparedness of a person for a meaningful perception of the battle and, oddly enough, an excess of ammunition. A soldier of the Red Army reached Berlin with a wearable ammunition load of 70 rounds for a mosquito. Experienced automatic riflemen during the Great Patriotic War said that they needed one disk for 71 rounds in a drum PPSh store. And it's not a shortage of ammunition. The more experienced and brave the warrior, the less often he shoots. And if it fires, then for sure. Shooting in combat must be meaningful. Short, 2-round bursts or single, while changing position.

36. Sight "3" and aim at the belt

The basis of marksmanship is the determination of the distance to the target. In the turmoil of the battle, there is no time to calculate the thousandth formula. Therefore, an experienced warrior sets the rear sight on the aiming bar of the machine gun at the number "3", which corresponds to a range of 300 meters, and aims at the middle of the opponent's figure. At the same time, at 100 meters of distance, the excess of the point of impact over the aiming point will be 25 cm for AKM, and 13 cm for AK-74 (this is for those who want to break the handle of a spoon). If this is too difficult for someone, feel free to put the rear sight on the aiming bar in the “P” position. It corresponds to the point-blank range (440m for AK-74, 350m for AKM) when the height of the bullet trajectory is not greater than the height of the chest target, and the shooter can fire without taking into account vertical corrections.

37. I don’t see, I don’t shoot

Constant moral and physical overstrain, fatigue, poor visibility - these factors of the battle seriously affect the psyche of a combatant. There is a desire to shoot at everything that can pose a threat. And then there’s the advice of the “experienced” - wet everything that moves, keep a cartridge constantly in the chamber, shoot at every rustle. The result of such a "science" are killed and wounded colleagues.

Usually they fire, without seeing the target. Untried beginners or finished cowards. If the former can still be taught, then the latter must be kept away. From them the threat comes no less than from the enemy. And also crazy shooting; a great opportunity for the enemy to sight our positions and inflict mortar or artillery strikes on them.

Therefore, the observance of the principle “I don’t see - I don’t shoot” should become the main rule of any fighter. Do not allow co-workers to hold weapons loaded, twirl them in different directions, or point them at a co- worker. Even as a joke. Maintain constant voice contact with co-workers. Don't shoot without seeing the target.

38. A few life hacks for combat shooting.

In case of an unexpected encounter with the enemy or in conditions of limited visibility - in the forest, in tall grass, un-mowed field - shoot low, at the level of the knee of the intended enemy. He will most likely lay down or crouch so as not to be a target. Fire in the building is best done from the depths of the room as the sound of the shot is muffled, and the flash becomes hardly noticeable. Approximately 90% of the world's population is right- handed. That's why shooting at the flash of a shot from the side of the enemy, move the aiming point half a figure to the right. In the window opening of the building, the shooter should be looked for in the lower right corner from you - because the natural shelter for him - the corner of the window and the wall, will be on the left from him.

Turning to the right is much more inconvenient to shoot than turning to the left. This is a psychophysiological feature of the body. Therefore, in battle, it is better to bypass the corners of buildings counterclockwise, staying away from these corners. With such a movement, the enemy loses the advantage. Staying in one place during the course of a gunfight without changing position means exposing yourself to additional risk.

It has been established that only a fifth of the personnel in battle act intelligently and conduct aimed fire. As a rule, these are commanders, snipers, and machine gunners. When such fighters are detected on enemy positions, they are destroyed by concentrated fire of all means of the unit. After that, the enemy loses combat stability.

39. Sometimes it happens ...

Even the most reliable weapons require maintenance and care. The main reasons for delays in firing a Kalashnikov assault rifle are a malfunction of the magazine and contamination of the rubbing parts of the weapon's mechanisms.

A faulty magazine is replaced by another, and dirty mechanisms are eliminated by their cleaning. Ideally, of course, clean the weapon after each shooting, which is unlikely in combat conditions. And no, you can't find gun lubricant RJ. Under these conditions, you can use kerosene, diesel fuel, WD-40 and even soapy water. The main thing then is to wipe the parts of the weapon with high quality cloth and dry, so that there are no traces of them left, which "grow" with dust and cause malfunctions. Do not lubricate the weapon with automotive, transmission, and even more so vegetable oils. Having thickened, they tightly "glue" the moving mechanisms of the weapon.

40. AK and weapon tuning

The ingenious designer Kalashnikov created a reliable and convenient machine gun, in which there is nothing superfluous that could cling, interfere or create discomfort for the owner. The machine is perfectly balanced for its design features. Therefore, tuning the machine gun with all sorts of devices has no functional significance, but is an elementary show off of people who are far from real participation in hostilities. Picatinny rails injure palms, clinging, tearing clothes, and when crawling, they contribute to contamination of weapons. A collimator sight, unlike a standard mechanical one, limits the view and does not allow the shooter to instantly transfer fire to another distance. On the 3rd-4th day, as a rule, it is simply lost, loosened in the fasteners. Expensive flame arresters and DTKs do nothing to improve accuracy, but they upset the balance of the weapon and make it heavier. Twin magazines connected with scotch tape or duct tape? You can recognize the headquarters or newcomers who want to look like tough fighters. In fact, twin stores are a sign of insecurity and even fear. They say that their owner has not been in a real battle and does not understand that such magazines carry the risks of contamination and breakage, upset the balance of the weapon, and make accurate shooting impossible.

41. Military majors against fighting bums

There is no universal military uniform. The main requirement for it is the strength of the fabric and ease of use. The fashion for military uniforms that has come to us with many pockets, zippers, patches, knee pads, etc. passes after the first two or three weeks of actual participation in hostilities. The most important thing when choosing clothes for war is that it does not differ from the uniform of your unit, and the army as a whole. Otherwise, there are great risks of falling under the friendly fire of their own colleagues who have accepted you for the opponent.

Experience suggested the need for year-round wear under uniforms of elastic moisture- wicking thermal underwear. It creates comfort, saves from scuffs, minor bruises, scratches, insect bites.

42. Sneakers, trekkers or boots

The special operation in Ukraine presented a number of special requirements for military footwear. Firstly, all-weather and waterproof. Secondly, tensile strength and reliable protection of the legs from injury. Thirdly, the sole must be solid, provide good grip, not be pierced by broken glass, rebar and sharp pieces of metal.

For these reasons, sneakers that have proven themselves in Afghanistan are not suitable for the Ukrainian theater of operations. Uniform high-top boots hold the ankle well and protect the legs from cuts and injuries.

This is where their advantages end. In the Ukrainian climate, they "die" after a couple of months of operation. Tactical trekking boots are lighter, more comfortable , have moisture-wicking properties and good grip on the ground. In such boots, the risks of rubbing bloody blisters on your feet are reduced. But the really cool "Lowa" and "Haix" are not available because of the price of 25-30 thousand rubles, and in muddy conditions they live no longer than state-owned boots.

In these conditions, it is worth thinking about cowhide boots and footcloths. By the way, during the winter Arden operation, the Germans, shod in wrapped boots, were ill with pneumonia four times less often than Americans who wore only boots.

43. A friend is a maid

Socks in boots "live" for two or three days. Therefore, in a situation where they have become unusable, and there are no spare ones, you can use footcloths, which are made of soft fabric. From sheets, shirts, T-shirts, etc. cut out 2 strips of fabric measuring 35 × 75 centimeters. It is important that they are without seams and scars that rub the legs. The footcloth fills more space in the shoe. In case of getting wet, it is easy to rewind it with the dry side on the foot, and the wet side on the shin, which minimizes abrasions during long walking. It is permissible to wear wet shoes on dry socks and footcloths, but not vice versa. During movement you can dry socks and footcloths, on the body, under the upper layer of clothing.

44. A soldier does not need extra property

No matter how weapons and military technologies change, the weight of equipment worn by a fighter of 30-35 kg remains stable for 150 years in almost all armies of the world.

What a fighter really needs to have with him:

- regular weapons and ammunition (12 kg);
- helmet and body armor (9 kg);
- unloading vest (1 kg)
- small infantry shovel (1.5 kg);
- water - 1.5 liters minimum;
- dry rations (2 daily allowances) (2 kg);
- raincoat-tent (1.5 kg);
- army bowler hat with a spoon (0.5 kg); - socks (2 pairs);
- first aid kit;
- toiletries (0.5 kg);
- uniform, boots, belt (3.5 kg);
- raid backpack (3.5 kg);

Total: 37 kg.

Resist the commanders demand that the fighter carry an additional ammo, 10 offensive and defensive grenades, ammunition for heavy weapons of the unit, etc. You can't fight with that weight. Therefore, the wise warrior carefully decides how to limit the load on the ridge. As a rule, to solve a combat mission not related to combat in the city, 4 magazines for a rifle and 2 grenades in unloading, 1.5 liters of water and one small infantry shovel for two are enough. Total, with the 15 kg. Bulletproof vest and helmet (9 kg) - optional.

The rest is stored in a backpack and transported in a combat vehicle. However, if the battle caught you on the march, in full gear - dump everything superfluous without regret. Retain weapons and ammunition. Collect the rest later.

45. 150 years in service

In the tsarist and Soviet Army, all the property of a soldier fit into a duffel bag made of tent fabric with a volume of about 30 liters. As 150 years of experience in its operation shows, nothing better can be invented. Brilliantly simple design. It is used as a backpack, as a filter for water purification, a head bag for accompanying a prisoner, a rakovka (?) and everything that a soldier's ingenuity is capable of. Total weight; 240 grams. The ideal useful volume allows you to accommodate everything that a fighter needs. A pea jacket and a sleeping bag are attached to the side straps, if necessary. Nothing extra should be shoved in.

The quilted straps of the duffel bag do not injure and at the same time do not interfere with aimed shooting with an emphasis on the shoulder. The bag is convenient as a support for shooting from a prone position. If necessary, it is not a pity to quit and act lightly. Not fashionable, not aesthetically pleasing, but cheap and practical.

46. Always keep with you...

The small infantry (sapper) shovel. It is a soldier's best friend

In a war, one cannot survive without a sapper shovel. A fighter must constantly dig in. The sapper shovel is an excellent tool for chopping blows. It can be used as an armor plate and even a frying pan for cooking. The main condition for the versatility of a shovel is a sharp blade. With a stupid shovel, as with stupid people, you won't get much. The lower and side ribs are sharpened on the MPL from the front (concave) side, the chamfer width is 3–5 mm.

A knife is an absolute must for everyone.

In 99 cases out of 100, you will need a knife only for cutting bread, sausage or peeling potatoes. Good, of course, if there is a multitool, but it will not replace a simple a kitchen knife with a blade up to 15 cm. And as for the fighters with "tactical" gut cutters sticking out for show from their loadbearing vest, it's like feathers in the tail of a peacock. More to impress.

Gloves - universal hand protection

Hands account for the majority of injuries and injuries, and any tactical gloves with protective elements will protect them from injuries, cuts, burns. Full-fingered gloves are better than fingerless ones, but fingerless ones are easier to shoot. If there are no tactical gloves, they, especially when digging trenches, will be successfully replaced by workers - 25 rubles per pair.

Important little things
"Knipser" is the correct name for nail clippers. Without care, nails will break, causing pain and discomfort. Crockery: bowler, spoon, mug (preferably stainless steel). Matches (preferably tourist, long burning), lighter. Scotch tape (skein). A very useful thing!

47. Without water neither there nor here
In winter, a person needs about 2 liters of water per day, and in the heat this rate rises to 4 liters. Lack of water causes rapid fatigue. Therefore, whenever possible, a soldier must stock up on water. 1.5 liters is an individual water NZ, which must be stored in every soldier's duffel bag. In a combat vehicle (usually in plastic canisters, tanks, etc.), a supply of water up to 100 liters per department (department - fighting unit).

Where to get water. To get started, study the map of the area where wells, springs, streams, lakes, rivers are marked. Such water is conditionally suitable for consumption, but after boiling or filtering, it is completely safe. In this case, it is good to have a portable filter for water purification in extreme conditions. (See in the Appendix: DIY Water Filter). In settlements, water must be sought in the fire department, in hospitals, and sanitary and epidemiological stations. There are special storage facilities for disinfected water. A decent amount of water after turning off the pumps remains in the water pipes in the basements of high-rise buildings. Especially located in the lowlands. Water should be consumed as thirst arises in small portions, no more than 100–150g. Anything more will immediately come out in the heat later.

48. War is war, but lunch is on schedule
Participation in hostilities is hard work, requiring huge energy expenditures of the body. Therefore, our dry food (IRP), depending on

the menu, contains from 4200 to 4700 kcal and provides the body with a sufficient amount of nutrients. But that's in theory.

In fact, there are two problems with dry rations. The first is that it doesn't exist. Second, it gets boring quickly. And dry ration weighs more than 2 kg and takes up a fair amount in a backpack. In Afghanistan, fighters, leaving for a raid in the mountains, for landing or for combing greenery, took the most energetically valuable from the dry ration - meat, lard, sugar, condensed milk. None of soups and cereals with compotes. At the same time, the weight was reduced three times, and the energy value - only 1.5 times.

Food obtained in the course of hostilities served as supplementary food. They ate, as a rule, twice a day—early in the morning and after sunset. We tried to diversify the diet and cook hot food (See in the Appendix: "Kulesh", "Unleavened wheat cakes", "Baked potatoes in foil on coals"). The optimal dish is kulesh. Thick soup from everything that is in the backpack plus cereals (any). Cakes are baked from flour. They are prepared for the entire squad - in the evening, in shelter, in disguise or in abandoned residents of private houses where there are stoves.

Hot food invigorates the body, digests much better and breaks down quickly without bloating and its accompanying problems - constipation or indigestion. In addition to gastronomic advantages, such catering contributes to team building, caring for comrades, and increasing the combat readiness of the unit.

49. Creeping enemies.

During the Second World War, 96 soldiers out of 100 suffered from pediculosis. The basis of the army of parasites were body lice. In addition to the discomfort of unbearable itching, they carry infectious diseases - "trench fever" and typhus. The number of victims of these diseases exceeded combat losses.

Existing effective sanitization products, as well as pest control chemicals, are often not available on the front lines. Our fighters have found an ingenious solution to this problem. In a clean barrel (large pan) of such height, so that the article placed in them does not touch the bottom or walls, two pieces of wood were inserted crosswise exactly along the diameter. After that, clothes were hung on them, under the tightly closed lid. The barrel was placed on stones and a fire was kindled under it. Half an hour later, in the hot barrel, all the parasites died. There are also known methods of dealing with lice, such as long-term boiling of the article, up to an hour, washing in gasoline, freezing (up to 12 hours in the refrigerator) and digging into the ground, in which a small edge is left at the top, on which lice are collected for subsequent destruction.

And one more detail. It is noticed that the advancing army suffers less from pediculosis. Active, resolute, cheerful fighters have almost no lice. But they can often be found in cowards and whiners.

50. Rules of the military "tenant"

In the conditions of total intelligence, which is conducted by our enemy, of course, is best placed in populated areas, on the territory of industrial enterprises with a system of basements and bomb shelters. Technique takes cove in workshops.

It is necessary to stay in the private sector in the 2nd or 3rd line of houses, while setting up guards 50–100 meters from the outskirts of the settlement. Choose to stay in a house with a basement and a source of water (well). It is better if the basement is taken outside the house and stands separately. It will be a reliable shelter in case of shelling. Do not stay in the most conspicuous and large houses. They are most likely shot or mined. Our scout Ilya Starinov destroyed German headquarters during the war years by mining such buildings. And today, Ukrainian rocket men are destroying ours where the chiefs did not study military history, and love luxury and comfort.

Do not set up a position next to a clearly defined landmark (corner of a grove, power line, a separate building, a house on the outskirts of a settlement, administrations , etc. This place will be hit by the very first artillery shot, because its coordinates are easily calculated from the map. If you happened to settle down in a field or steppe, then do not be lazy here. Occupy the main position and a couple of false ones, into which you dump garbage and household waste, make a fire there if you need to - in short, you demonstrate active habitation. On the main one, there is a minimum of movement. Opening up 2-3 company positions is hellish work, but this is a sure way to deceive the enemy's reconnaissance and artillery in order to survive. In all cases, as experience shows, it is better to withdraw military equipment from positions so as not to provoke artillery fire and shelter no less than a kilometer from the bases position, providing 2-3 routes of advancement to it.

51. Dug in - remained alive

With an artillery attack, the chances of surviving a soldier in a trench are 10 times higher than that of an unentrenched one, and in connection with the massive use of long-range weapons by the enemy, the concepts of front and rear are blurred. Danger is everywhere now. Therefore, those who really fought do not consider it shameful to dig a trench for themselves. At a halt during the march, on the front line, when it is not possible to build a strong point for the unit, in the places of deployment of the second echelon - as a rule, trenches are dug at least 1 meter deep and 80–90 cm wide so that two people can sleep side by side, warming each other, and the raincoat tent is wide enough to protect from the weather. Coniferous branches, straw, etc., laid on the bottom, will save you from dampness. If conditions permit, the trench is covered with branches, slate, plywood, sprinkled with soil, which serves as a shelter from small fragments. In the cold season, in such a trench, you can place a "Polaris" or a mini rocket stove ("Swedish candle - the simplest hearth", "Polaris", "Rocket stove"). If there is time, the trench is equipped with a "fox hole" - a niche in the side wall facing the enemy. Sometimes, when equipping platoon and company strongholds, commanders involve equipment that is not adapted for such work. Most often, road graders / diggers. As a result, instead of a narrow trench that protects the fighters from bullets and shrapnel, a crooked pit with crumbling edges is obtained. It does not save from artillery fire, it does not cover the quadrocopter, and in bad weather it turns into a muddy pond. Not a shelter, but a trap for his own unit.

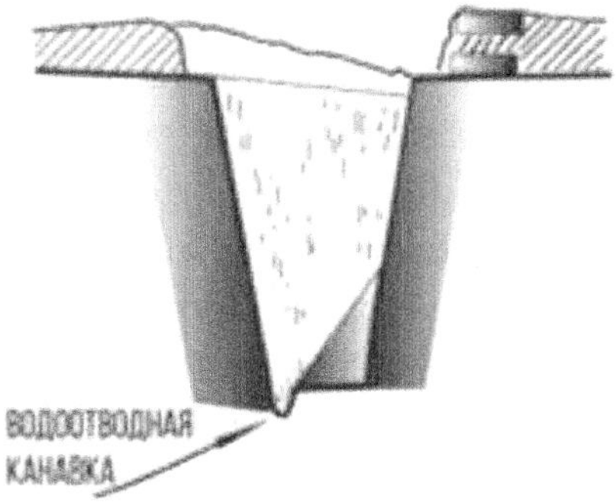

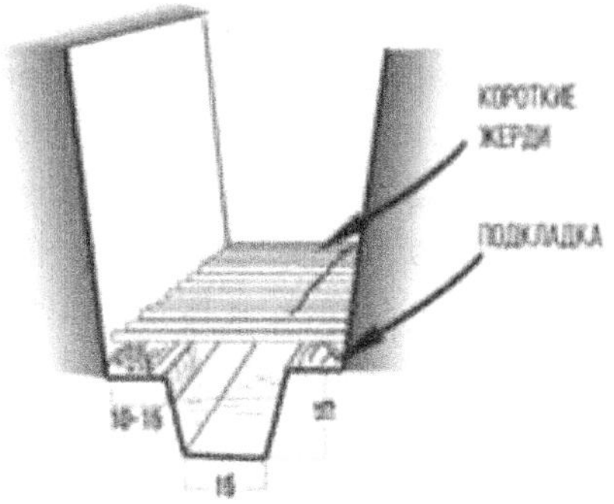

Fig. 1. Drainage ditch of trench Fig. 2. Overlapped drainage bottom ditch at the bottom of the trench

Another pitfall during the construction of the supporting structures is associated with the lack of a drainage system at the bottom of the trench. Without it, trenches in the rain fill with water and mud, which create discomfort for the inhabitants. A drainage ditch to the depth of the shovel / bayonet (20 cm) is dug out for the entire length of the trench at its back side and goes to the lowest point of the support, where a catchment well is arranged (See Fig. 1). The damp and viscous bottom of the trench is covered with gravel, sand or straw. In the presence of material time and material water a diversion ditch is arranged in the middle of the trench, overlapping it with a deck of boards or poles (See Fig. 2).

52. Trench—5 stars

Where the fighting has acquired a positional character, and the fighters are forced to live for months in damp trenches, if you are not lazy, you can get comfortable. To do this, abandoned vehicles are buried in the ground at positions - a minibus, a passenger station wagon, a bread truck booth. Everything will fit. Seats are thrown out of the cabin, the floor is leveled with flooring and sleeping places of increased comfort are ready. Precipitation, cold, shedding of earth and groundwater is nothing to this “spruce up”. And if the trench is well deepened, disguised and securely covered from above, then it can be considered as a dugout of an increased comfort class.

53. Camping out of need

Under these conditions, constant reconnaissance by the enemy for targets using drones and mining of the routes of advancement of our troops turn a campaign of necessity into a kind of quest with an unpredictable result.

The traces of this, so to speak, activity, scattered by the wind across the fields and plantings, unmask defensive positions, and in order not to "fly in" at the most inconvenient moment, it is necessary to hide the latrines as carefully as possible. It is better to organize them in enclosed spaces - abandoned houses, workshops, farms. In positions located outside settlements, a latrine is arranged 30–40 m behind the front trench and connected by a communication course. The length of the ditch of the latrine is determined at the rate of 2–3 m per compartment. Boards are used for its arrangement or shipping pallets with sawn-off versions (See fig. 3 and fig. 4).

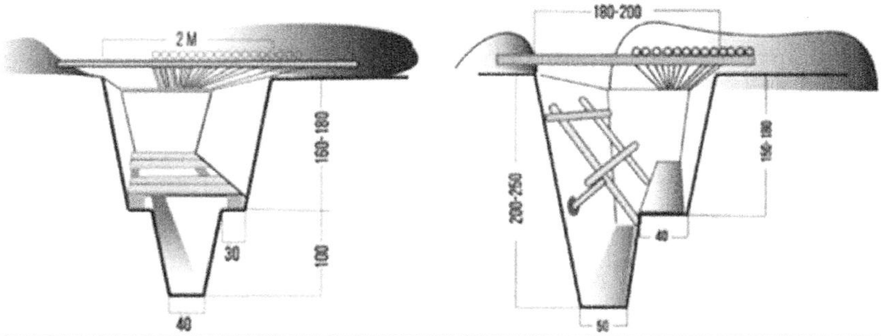

Fig. 3. Lavatory in the trench, Fig. 4. A latrine in a trench equipped with a pallet equipped with poles.

It is even more difficult to tend to natural needs on the march, or on the offensive. A latrine trip is fraught with risks of running into a trip wire or an enemy DRG. No matter how ridiculous it may sound, but in Afghanistan, for example, it was customary to cover a comrade in this situation. Veterans who went through the Great Patriotic War said that it was important to accustom the body to the toilet in the morning before dawn. And for this it is not necessary to devour the entire dry ration during the day and supplement it with products found in abandoned houses. And in general, before the fight it is better to have an empty stomach: with a wound in the abdomen, there is less chance of peritonitis.

54. Sign. Garbage in position brings shelling

The enemy chooses to strike the weakest link in our defense. Artillery strikes are inflicted on it, it is stormed by enemy infantry and gutted by reconnaissance. Calculating the weak link is easy. The amount of trash in the position reveals not only the number of defenders, but also the nature of the organization of the service. Garbage is a direct result of low discipline. The mess in the trenches - linen on the ropes, rubbish on the parapets, the filthiness of the plantings and fields around the position due to the lack of a latrine, speaks of the absence of a firm commanding hand and the uncontrollability of the unit, where everyone lives their own lives. And this is the weak link, which will definitely be hit.

If you are unlucky with the commander, then the fighters themselves must protect themselves. To do this, you need to ensure that garbage is always covered in positions. In niches, equipped covered with plank covers, if the trenches are located outside the settlement. In a garage closed from the wind, a building, a basement - if in a city or village. So that the orderlies daily sprinkle the latrines with earth. This will rid the position of unpleasant odors, diseased microorganisms, birds, rats, and other living creatures. So that each fighter works daily to strengthen the trenches and turn the position into an impregnable fortress. Only in this case the enemy will not dare to storm it, realizing that he will deal with a strong spirited garrison.

55. Aibolit himself

Our body is a self-organizing system and the higher the voltage, the more internal reserves it connects to our protection. Even in the most difficult conditions of trench warfare - in the cold, in clothes that are wet and freezing on the body, the fighters practically do not get sick. Therefore, it makes no sense to keep a warehouse of medications with you. But all kinds of infections and boils, entailing suppuration, dysentery and eczema, really cause a lot of problems. Therefore, at every opportunity, you should wash your clothes, wash your feet, hands, groin, armpits, and dry your shoes.

What is really useful to a fighter and should always be at hand:

- hemostatic tourniquet; PP
- individual dressing bag or bandage 10 cm x 5 m;
- a set of bactericidal adhesive plasters (Bandaids) of various sizes and an adhesive plaster in a roll;
- 2-3 gauze or cotton pads.
- Of the tablets - loperamide - a remedy for diarrhea.

Two tablets at once, plus one after each emptying, but not more than four per day. There is a risk of getting the opposite effect. For allergy sufferers - antihistamines prescribed by a doctor. From the middle of summer until frost in Ukraine, the strongest allergen, the ragweed grass, actively blooms. As well as drugs prescribed by a doctor on a regular basis.

56. Here the bullet flew by and—yeah...

Rapid bleeding and pain shock are the cause of death in half of those who died from gunshot wounds. 30% of such wounded die within the first hour after being wounded. These are medical statistics. It can be changed, saving the lives of our fighters, by learning how to properly provide first aid to the wounded. Training in this matter is useless. You will not become a military field surgeon anyway.

Therefore, in an extreme situation, when injured, two requirements must be met: First, if a wounded person is bleeding in front of you, we inject Promedol (opioid). To do this, with a strong movement, at a right angle, insert the tube needle for its entire length into the soft tissues of the buttock, thigh or shoulder, squeeze out the entire contents of the syringe and remove the needle.

Second, the wounded limbs are pulled with a tourniquet with all the force, to the maximum, until the bleeding stops and after that we bandage the wound. Indicate on the tourniquet the time of its application. You can not keep the tourniquet for more than two hours. It is necessary to weaken and drag above the first place of imposition.

For a wounded person in the chest or abdomen, after an injection of Promedol, we plug the wound with a bandage, cotton wool, cloth, IPP wrapper and bandage it tightly. We transfer all the wounded as soon as possible to the physicians.

57. Anesthetize and do no harm

Promedol is an anesthetic used for wounds and injuries, a narcotic analgesic. The action of Promedol occurs within 10-20 minutes and continues after a single dose for 3-4 hours. It makes no sense to inject more than one tube of Promedol when injured. The pain-relieving effect is not enhanced, but the wounded may die. The drug depresses the respiratory center.

After the injection of Promedol, in a conspicuous place on the wounded - on the form, on the face, chest, write the time of injection. If you are injured in the head, you cannot inject Promedol.

58. Being captured is a bad, very bad idea.

They are captured for various reasons. The wounded, unconscious, left without ammunition and commanders. But most often out of stupidity - he got lost, went to the village for cigarettes, fell asleep at the post. Sometimes voluntary surrender, but this is a matter of psychiatry. The one who considers captivity as a way to save his life is mistaken. Being captured is a very bad idea. During the Great Patriotic War, 63.4% of Soviet prisoners of war died of starvation, disease, and torture. More than half of the prisoners died in Afghan captivity. Ukrainian warriors also arrange hell for our military prisoners of war.

Behind the bestial cruelty towards our captives hides the awareness of inferiority and the enemy's fear of the Russians. And for them we are all - Chechens, Tatars, Buryats, Abkhazians, Udmurts - Russians. To humiliate a defenseless prisoner, to torture, castrate, kill - for our enemies is a kind of act of self-affirmation. And for us- an unacceptable zapadlo. We don't torture prisoners. Save the life of the surrendered to the enemy - a charitable and legally correct deed.

59. There are no hopeless situations

Anything happens in a war. You can meet an enemy that is superior in strength, get surrounded, or be left without ammunition. The main thing in such a situation is not to panic.

If you don't know what to do, drink some water. This will take the pressure off. Choose a safe hiding place and lie down. For this, a forest, planting or industrial zone is best suited. Through them, then it is best to break out of the environment.

Try to wait until dusk. Firstly, against a nickname in the twilight it is difficult to figure out where your friend is, where someone else's. Secondly, an open sight on a machine gun is no longer effective, a night sight is not yet used so as not to illuminate the matrix, thermal imaging sights are still rare, which means that shooting at you will not be as effective as during the day. Often twilight is accompanied by fog, and this too helps you stay undetected.

If found, change cover as soon as possible. It is difficult to hit a running person even at a short distance. If the enemy is pursuing, and you have no strength left to break away, organize an ambush on him. Let go and shoot point- blank. Burdened with the wounded and dead, the pursuers will lose all interest in you. Going out on your own is better during the day. Less risk of falling under "friendly fire". Until then, even if you meet our fighters, stay in cover. Making sure that you see your own, attract their attention, and make it clear that you are not an enemy. Only then do you come out of hiding. Do not give up! Never!

60. Don't drink alcohol, you'll become a blind man

Nothing, perhaps, did the army more harm than the legends about the benefits of the people's commissar's 100 grams. A drunk person will either get hurt or killed - that's no question. Drunken noise will attract the attention of the enemy, and he will throw shells into the trenches or send the DRG to you.

For the sniper a drunken fighter is an easy target. A drunk, without understanding, can open fire on his own. And also stupid orders, hazing. All this is alcohol. If you can't stop your colleagues from drinking, leave. It doesn't end well. The only thing vodka is good for in war is for rubbing the body when there is no way to bathe and for softening the skin of boots. The ankle boots are abundantly moistened from the inside with vodka and worn on the leg until dry.

61. Looting

Bakshish in Eastern languages literally means a gift. More broadly, a trophy. Money, jewelry, all kinds of gadgets, cars - everything that was of value in civilian life is worth absolutely nothing in war. The main value in war is life. And in order to preserve it, one must give up all thoughts of material things and live in war. It has been noticed that as soon as a fighter begins to roam through abandoned houses, consider that he signed his own sentence. War does not tolerate marauders and cleans them up first. Be sure that the bakshishniks (One who seeks out bakshish) will collect all the mines and booby traps in the area, come under friendly fire from the grays, reveal your positions, and expose themselves and you under enemy artillery.

But the most vile thing is that bakshish changes people. Cheerful chaka, they turn the hero into a coward and a miser who lives only for the sake of his trophy. Such a fighter becomes a burden. It's of no use. Around him only conflicts and quarrels. You must stay away from him. According to the law of karma, it is in his trench that the largest shell will fly, the most accurate bullet, or the most corrosive investigator will come. No one has canceled the criminal offense for looting.

62. War will not write off

If someone thinks that one can “relax” in war without observing the written laws and unwritten customs of war, then he is mistaken. The first months of confusion and confusion in the NWO are over. The military police and counterintelligence agencies were at full strength. All crimes committed by military personnel will be disclosed without fail.

Almost 80 years have passed since the end of the Great Patriotic War, and investigators are still looking for and finding war criminals. Today, in the era of information technology and total digital control, it is impossible to hide the facts of murder, violence, and robbery. Russian special services, as practice shows, get all sorts of "shaitans", wherever they hide. This applies to all those who have committed crimes, both the VSE units with mercenaries and our servicemen, regardless of their rank and position.

63. Don't pee, brother!

In fact, fear is just a reaction of the body to danger; within reasonable limits, it helps to avoid a lot of trouble. Fear, for example, of an artillery attack makes us dig deeper into the ground. Fear of the sniper prompts you to move in dashes.

Fear goes away as we gain combat experience, when we understand that not every bullet is in the forehead, and not every projectile is in your trench. So it turns out that fear in war is a derivative of the lack of information about the enemy and its possibilities. The easiest way to overcome your fear is to be in a team, next to experienced fighters, when there is an opportunity, speak out. Friendly - "Don't piss, brother!" works better in combat than any tranquilizer. Emotional relaxation also helps a lot - obscenity towards the enemy. But the first remedy for fear is the offensive actions of our army. When you see with your own eyes the broken enemy equipment, its crushed positions, squinted or captured ukrovoyaks, miserable, exhausted, and dirty, it comes to understanding that the devil is not really as terrible as he is painted.

64. The most important thing. God with us!

“God leads us, He is our general!” - wrote the great Suvorov in his book "The Science of Victory". Sincere faith in God, the primacy of the spiritual over the material - this is the main secret of the victories of the commander and his miracle god entire. “Neither hands, nor feet, nor the mortal human body wins, but the immortal soul, which rules with both hands, feet, and weapons, ...”, “Without prayer, do not draw weapons, do not load guns, do not start anything!” , "Pray to GOD:

victory comes from HIM!" Suvorov teaches us. (See in the Appendix "Prayer of Orthodox Warriors Before the Battle", "Troparion to the Cross and Prayer for the Fatherland", "Du'a of Faithful Muslims, Which Helps to Win"). And Alexander Vasilievich firmly knew that there was no death, and death in battle was just a stage on our path to God. The test of war is a kind of purgatory, after which we purify our souls and gain faith, and God's providence is to accept our souls, or leave them on Earth for only He alone knows the future.

65. We will go to heaven, and they will just die

The fact that Russia and we, its soldiers, are fighting today on the side of God is evidenced by the fact that thousands of Russians of all nationalities and different religions, without coercion, at the call of conscience, became soldiers in order to fight evil. This happened only once in history - during the Great Patriotic War, where our grandfathers fought against raging fascism.

Today we are Orthodox and Muslims, disty and shamanists in the same formation fighting against Ukrainian nationalism and the world behind them of Satanism.

And, perhaps, our President was not joking when he said that "... we, as martyrs, will go to heaven, and they will simply die." Probably, Putin knew something that he didn't have time to say.

66. If the pamphlet goes to the enemy

Let them read. Sooner or later Ukrainians will become Russians again, because they have always been Russians. And the more current fighters of the Armed Forces of Ukraine save their lives, the fewer tears their loved ones will shed. The special military operation will end, and the current Ukrainian servicemen will once again be shoulder to shoulder with the Russian soldiers to resist the West, which provoked this fratricidal war. The Romanians and Finns fought together with the Germans for three years against the USSR, and then, as soon as the boot of a Soviet soldier set foot on their land, they turned their guns against the Nazis. It has always been so. So it will be this time.

Prayers

Troparion to the Cross and Prayer for the Fatherland

Save, Lord, Thy people, and bless us
Yours, granting victory to the opposing ones, and I keep Thy dwelling place by the Cross.
Translation: Save, O Lord, Thy people and bless those who belong to Thee, helping to overcome enemies and preserving Thy holy Church by the power of Thy Cross.

Prayer of Orthodox soldiers before the battle

"My Savior! You laid down Your life for us to save us. You commanded us to lay down our lives for our friends, for those close to us. Joyfully I go to fulfill Your holy will and lay down my life for the Tsar and the Fatherland . Arm me with strength and courage to overcome our enemies and grant me to die with firm faith and the hope of eternal blessed life in Your kingdom. Holy Mother of God, save me under your roof."

Du'a of true Muslims, which helps to win

This is the Messenger of Allah (peace and blessings of the Almighty be upon him) personally entering into the fight with enemies

ipe ipe ipe ÿÿÿ .ipe .ipeÿuzzÿ icy icy icy ÿleb, ÿÿakeÿ ipe ipe ipe faceuleÿ islamoring miss .
"Allahumma munzil al-kitab sarig al-hisab, ih
zim il-ahzab. Allahhumma-hzimhum wa zalzilhum."

Oh, God, who sent down to us the Noble Koran, before which we have to answer.
Help win over the enemy. Scatter and overthrow them from society. (Nawawi, al-Adhkar: 188).

Made in the USA
Monee, IL
25 March 2025

14602551R00030